Amazing Animals with Flowers Coloring Book

Cool, Creative and Original Animals Drawings for Toddlers and Kids

The possession of this book belongs to:

From the author:
(Syleray and Sixela)

As a lover of nature and animals, I created this book in hopes of inspiring children to explore and learn about the amazing animal world through art and color.

Each design has been carefully crafted to capture the beauty, tenderness and diversity of some of the animals that share our planet and home.
I hope you enjoy coloring and learning with this book as much as I enjoyed creating it.

Let your imagination run wild!

If you want to get in touch with me, scan this code. Feel free to follow me and stay tuned for new posts.

HI THERE, MY COLORFUL FRIEND!

COLORING CAN BE SUPER FUN, AND IT ALSO HAS SOME COOL BENEFITS:

- Express Yourself: Coloring lets you show your creativity and feelings through colors and shapes. You can make your world more colorful!
- Relax Your Mind: It's like a mini vacation for your brain. Coloring helps you chill out, like taking a deep breath for your thoughts.
- Learn and Improve: Coloring can help you get better at staying inside lines and using different colors. It's like practice for being neat and learning about colors.

EXCELLENT TIPS TO ENSURE THE BEST COLORING EXPERIENCE:

- Use quality coloring supplies.
- Color in good lighting
- Plan your colors, but be creative and patient.
- Stay within the lines to create beautifully detailed illustrations.

Remember!

All great things need time, perseverance, and patience.
This coloring book is designed for all coloring skill levels.

HAPPY COLORING!

Page summoned to test hues.

Thank You!

We hope you enjoyed this book.?

Your feedback
is incredibly valuable to us.

WE DON'T EXIST WITHOUT YOU!

Show us your drawings on Amazon and leave us your
Review

Inner and cover graphics designed by MidJourney and Canva
Composition of elements: Syleray & Sixela